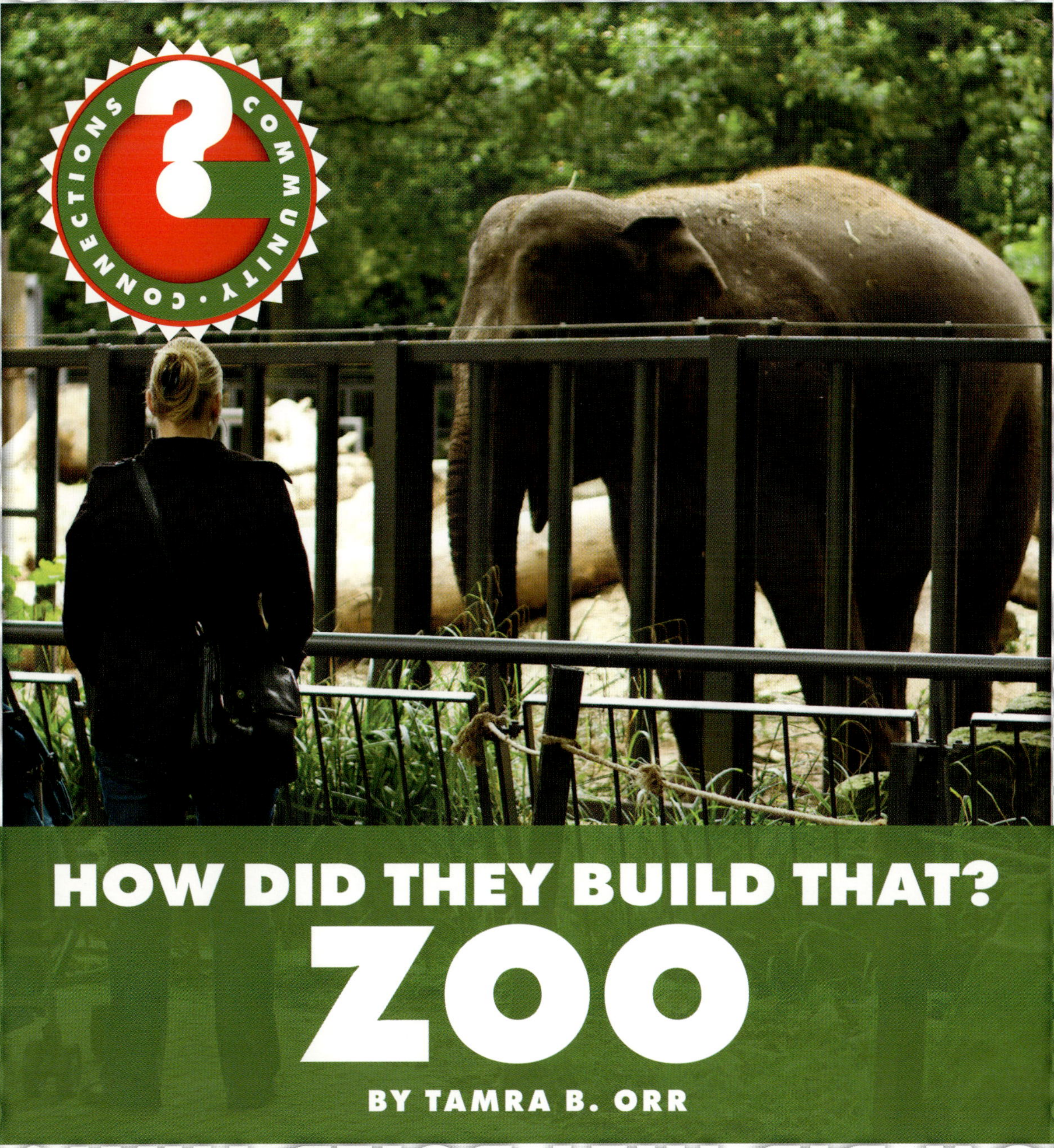

HOW DID THEY BUILD THAT?

ZOO

BY TAMRA B. ORR

Published in the United States of America by Cherry Lake Publishing
Ann Arbor, Michigan
www.cherrylakepublishing.com

Content Adviser: Susan A. Chin, Vice President of Planning and Design and Chief Architect, Wildlife Conservation Society, New York

Photo Credits: Cover and page 1, ©Linda Kennedy/Alamy; page 5, ©Chuck Eckert/Alamy; page 7, ©Asia Glab/Shutterstock, Inc.; page 9, ©Karen Givens/Shutterstock, Inc.; page 11, ©freedomman/Shutterstock, Inc.; pages 13 and 15, Julie Larsen Maher ©Wildlife Conservation Society; page 17, ©Len Holsborg/Alamy; page 19, ©iStockphoto.com/Adventure_Photo; page 21, ©istockphoto.com/Image_Source

LIBRARY OF CONGRESS CATALOGING-IN-PUBLICATION DATA
Orr, Tamra.
How did they build that? Zoo/by Tamra B. Orr.
p. cm.—(Community connections)
Includes bibliographical references and index.
ISBN-13: 978-1-61080-113-3 (lib. bdg.)
ISBN-10: 1-61080-113-X (lib. bdg.)
1. Zoos—Juvenile literature. I. Title.
QL76.O77 2011
590.7'3—dc22 2010052869

Cherry Lake Publishing would like to acknowledge the work of The Partnership for 21st Century Skills. Please visit *www.21stcenturyskills.org* for more information.

Printed in the United States of America
Corporate Graphics Inc.
July 2011
CLFA09

ZOO

CONTENTS

LET'S BUILD A ZOO!

Have you ever been to a zoo? They are fun to visit. There are many animals to see.

Monkeys climb and play in their **enclosures**. Reptiles slither and slide. Big cats roar and prowl.

Zoos let us see animals we might not get a chance to see in the wild.

BROOKFIELD ZOO
Chicago Zoological Society
Inspiring Conservation Leadership

It takes a lot of planning to build a zoo. Zoo planners decide how many animals will live in the zoo. Then they figure out how much space each animal will need.

Some animals need indoor space. Others live outside.

Some zoos are home to rare animals such as these pandas.

LOOK!

Look around the next time you go to the zoo. What do you notice about the enclosures? Which ones have plants? Which ones have ponds? What does this tell you about the animals that live there?

WHO WILL LIVE HERE?

It is important to decide which animals a zoo will have. There are rules about which animals can live in zoos. Zoos get animals by sharing and borrowing from other zoos.

Each **species** needs the right enclosure. Enclosures need to be a lot like the animals' natural **habitats**.

Enclosures are made to look like the places where animals live in the wild.

Zoo enclosures need certain things to keep animals happy. Each animal likes different kinds of plants. Some animals need pools to swim in. Others need a lot of space to run around.

Some animals need enclosures that are very cold. Others need a lot of sunlight.

Gorillas need things to climb on in their enclosures.

THINK!

Enclosures must keep animals from escaping. What does this mean for animals that can climb, fly, or jump? What kind of walls could keep them in? What shapes do their enclosures need to be?

DON'T FORGET THE VISITORS!

Zoos also need to keep visitors happy. People need benches and chairs to rest on. They need parking lots for their cars.

Walkways help visitors move around the zoo. Shops give them places to buy snacks and gifts. Don't forget that zoo visitors also need restrooms!

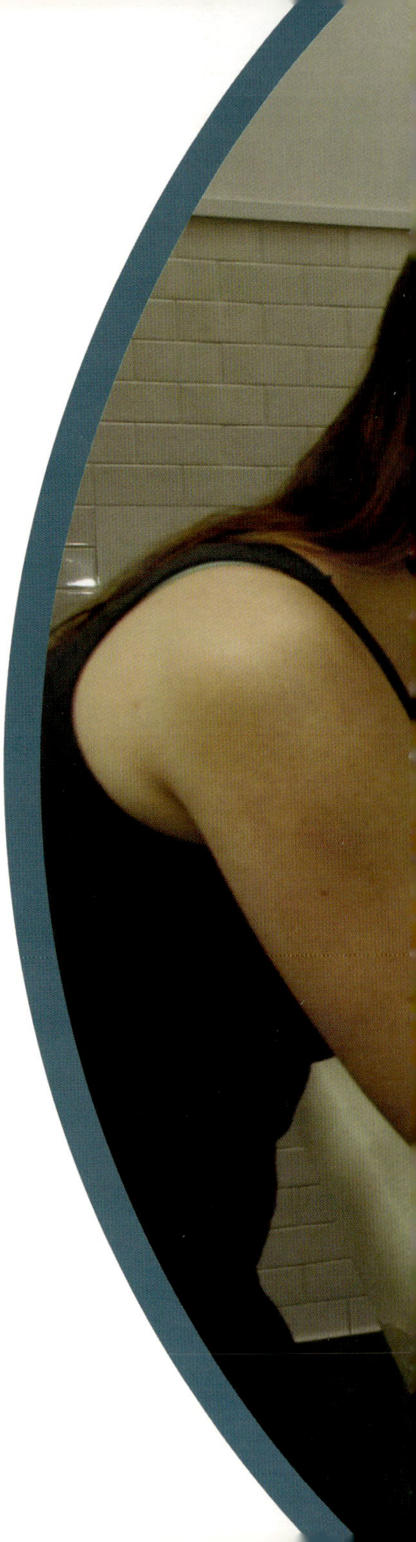

Zoo restrooms often have water-saving faucets and other eco-friendly features.

dinosaurs!) The existing
keeps recirculating, which is
why we must conserve it.
Water, Water Everywhere?
You Can Save Water
Keep water in the fridge instead of running the tap for cold water. Over two gallons of water flow from the tap in just one minute!
Water from this sink will be a plant's drink! This water will go down the sink to the gray water garden outside.

Zoo workers need space, too. They need places to take breaks. They also need places to store their things while they work.

Veterinarians need room to care for sick animals. Other zoo workers need offices with phones and computers.

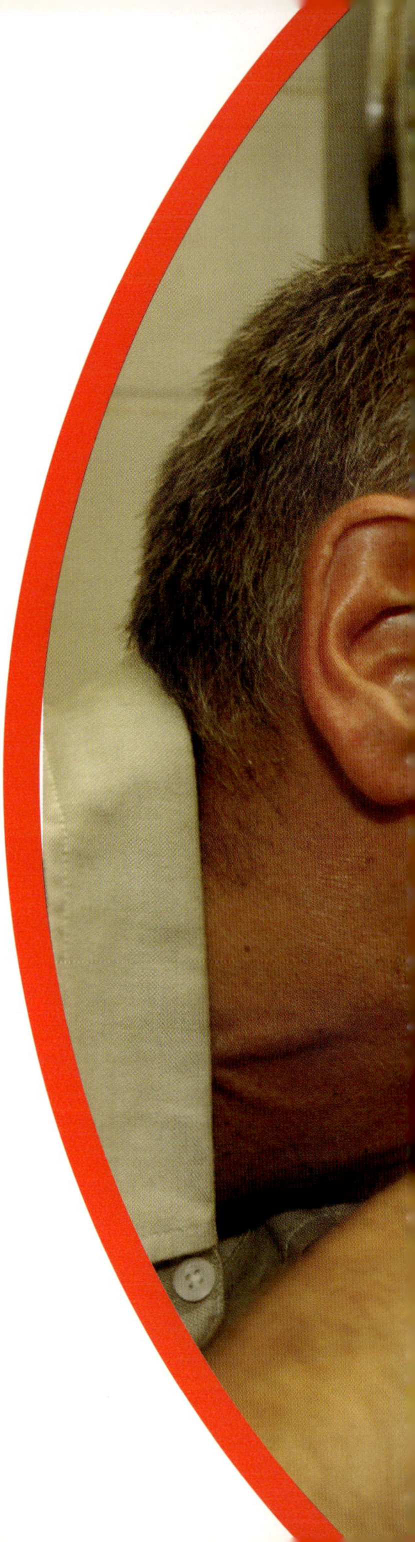

Veterinarians need offices for taking care of zoo animals.

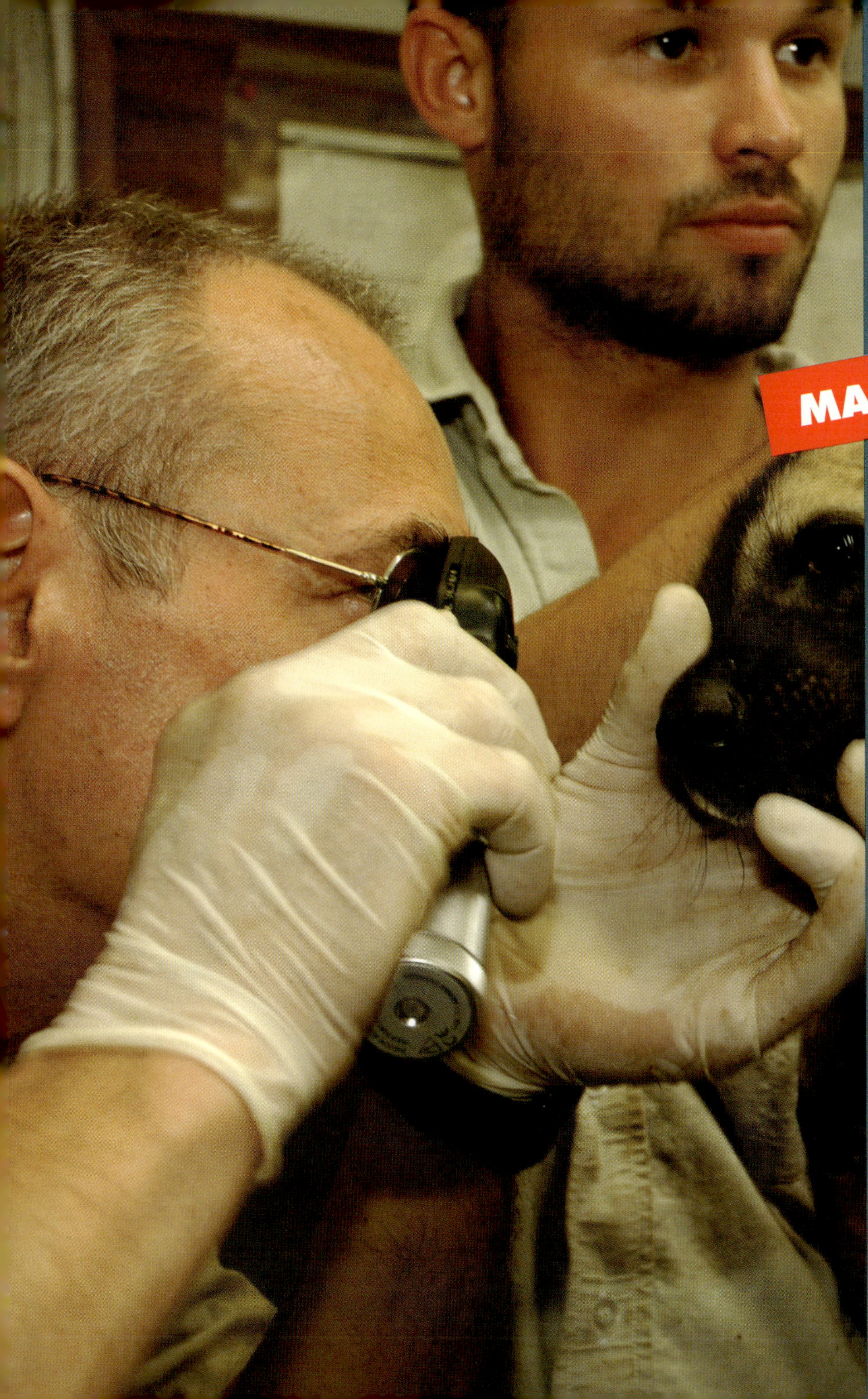

MAKE A GUESS!

What would it be like to be a zoo veterinarian? What kinds of animals would you work with? Which injuries would be common? What would you do to keep the animals healthy?

Zoos are very busy places. Zoo planners need to make sure there is room for all of the people and animals. Most zoos are built on large pieces of land.

Zoo planners also need to make sure the zoo is built where people can get to it easily. No one will visit if the zoo is too far away!

Zoo planners have to make sure that visitors can easily find their way around the zoo.

THE BRONX ZOO
Store
Bronx Zoo
Popcorn
5

PUTTING IT ALL TOGETHER

Construction takes many months. Dozens of builders work hard to make sure the enclosures meet the needs of the animals. A single enclosure can cost millions of dollars!

Big animals, such as giraffes, need enclosures with room to move around.

The zoo is a great place to learn about animals. Now you know how much hard work it takes to build one.

What would you include if you could build your own zoo? Your ideas could help improve the zoos of the future!

Some zoos let you get a close-up look at certain animals.

CREATE!

Have you ever been to a petting zoo? These special zoos let you touch the animals. Make a list of animals you think would be good for a petting zoo. What animals would not work? Why?

GLOSSARY

construction (kuhn-STRUHK-shun) the process of building something

enclosures (en-KLOH-zhurz) fenced-in areas of the zoo that hold animals

habitats (HAB-uh-tats) the environments where creatures normally live in the wild

species (SPEE-sheez) a grouping of animals that are alike and can produce young together

veterinarians (vet-er-uh-NEHR-ee-uhnz) doctors who work with animals

walkways (WAWK-wayz) paths for walking

FIND OUT MORE

BOOKS

Jennings, Terry. *Mammal Kingdom*. Mankato, MN: QED Publishing, 2011.

Orr, Tamra. *Working at the Zoo*. Ann Arbor, MI: Cherry Lake Publishing, 2011.

WEB SITES

Wildlife Conservation Society

www.wcs.org
Learn more about what can be done to protect wildlife around the world.

Smithsonian Friends of the National Zoo: Just for Kids

nationalzoo.si.edu/Audiences/kids/default.cfm
Learn fun facts about zoo animals and read stories about zoo veterinarians.

INDEX

ABOUT THE AUTHOR

Tamra B. Orr is the author of more than 250 books for readers of all ages. She and her husband live in the Pacific Northwest, and she is mother to four, all of whom have made sure she has gone to the zoo.